Kiran Gangarapu

Development if Isatin as **CNS** Agents

Anticonvulsant activity

Anchor Compact

Gangarapu, Kiran: Development if Isatin as CNS Agents: Anticonvulsant activity. Hamburg, Anchor Academic Publishing 2014
Original title of the thesis: SYNTHESIS OF NEW ISATIN DERIVATIVES FOR POSSIBLE CNS ACTIVTIY

Buch-ISBN: 978-3-95489-279-2
PDF-eBook-ISBN: 978-3-95489-779-7
Druck/Herstellung: Anchor Academic Publishing, Hamburg, 2014

Bibliografische Information der Deutschen Nationalbibliothek:
Die Deutsche Nationalbibliothek verzeichnet diese Publikation in der Deutschen Nationalbibliografie; detaillierte bibliografische Daten sind im Internet über http://dnb.d-nb.de abrufbar

Bibliographical Information of the German National Library:
The German National Library lists this publication in the German National Bibliography. Detailed bibliographic data can be found at: http://dnb.d-nb.de

© Anchor Academic Publishing, ein Imprint der Diplomica® Verlag GmbH
http://www.diplom.de, Hamburg 2014
Printed in Germany

CONTENTS

CHAPTER-1

CHEMISTRY OF ISATIN

INTRODUCTION

Isatin (1H-indole-2,3-dione) (I) was first discovered by Erdmann[1] and Laurent[2] in 1841, independently as a product from oxidation of indigo by nitric and chromic acids.

(I)   **(II)**

It is a unique molecule possessing both amide and keto carbonyl groups. Apart from this, it has an active hydrogen atom attached to nitrogen (or oxygen) and an aromatic ring which was substituted at 5- and 7-positions. It exists in a tautomeric form (II) and these functional characteristics play an important role in governing the various reactions of the molecule.

The C-3 carbonyl group of isatin is strongly electrophilic. As a result, isatins are readily involved in condensation and addition reactions with carbanion type nucleophiles into 3-substituted oxindoles[3]. In general, there are three possibilities during condensation reactions.

i)   both the α, β-carbonyl groups, having varying in reactivity are involved,

ii)   ring cleavage takes place and

iii) ring expansion occurs

A general observation reveals that the nature of final product always depends on the experimental conditions and substituents on nitrogen atom which may affect the electron density at α and β carbonyl carbon atoms respectively[4].

Synthesis of Isatins
The Sandmeyer methodology :

The method developed by Sandmeyer is the oldest and the most frequently used for the synthesis of isatin. It consists of, reaction of aniline with chloral hydrate and hydroxylamine hydrochloride in aqueous sodium sulfate to form an isonitrosoacetanilide, which after isolation, when treated with concentrated sulfuric acid, furnishes isatin (1) in >75% overall yield (Guo and Chen, 1986). This method is applied mostly well to anilines with electron-withdrawing substituents, such as 2-fluoroaniline[5].

Scheme-1

In addition to the use of H_2SO_4 for the cyclization step, isonitrosoacetanilides is heated in $BF_3.Et_2O$ at 90°C. After cooling the reaction mixture, addition of water allows isolation of the respective isatins. This methodology has been proved to be particularly effective for the preparation of benzo-oxygenated isatin derivatives[6,7].

The Stolle Procedure

The most important alternative to Sandmeyer's procedure is the method of Stolle. In this method anilines react with oxalyl chloride to form an intermediate chloro-oxalylanilide which can be cyclized in the presence of a Lewis acid, usually aluminium chloride or $BF_3.Et_2O$, although $TiCl_4$ has also been used to give the corresponding isatin. This method has been used for the synthesis of 1-aryl[8] and polycyclic isatins derived from phenoxazine, phenothiazine and dibenzoazepine[9] as well as indoline[10]. In the case of dimethoxyanilines, spontaneous cyclization to yield dimethoxyisatins in the absence of a Lewis acid has been observed, as exemplified in the synthesis of melosatin A (2), albeit in very low yield (Scheme-2).

Scheme-2

The Martinet Isatin Synthesis

The Martinet procedure for the synthesis of indole-2,3-diones involves the reaction of an aminoaromatic compound and either an oxomalonate ester or its hydrate in the presence of an acid to yield a 3-(3-hydroxy-2-oxindole)carboxylic acid derivative which after oxidative decarboxylation yields the respective isatin (3). This method was applied successfully for the synthesis of 5,6-dimethoxyisatin from 4-aminoveratrole whereas the use of 2,4-dimethoxyaniline was less successful[11] (Scheme-3).

3

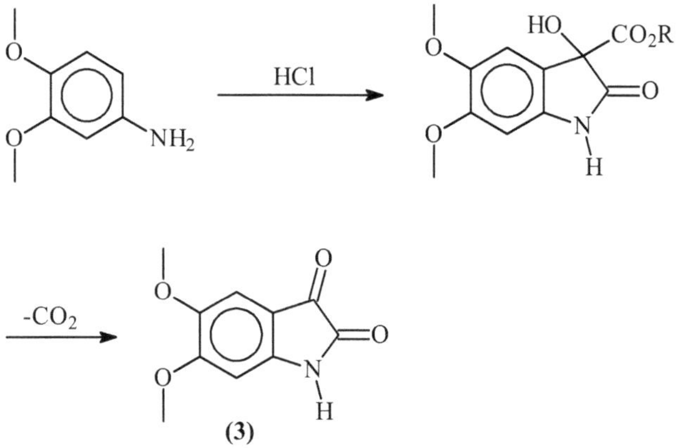

Scheme-3

The Martinet procedure is readily applied to naphthylamines, thus yielding benzoisatin derivatives[12].

The Gassman Procedure

A fundamentally different and general procedure developed by Gassman is another option for the synthesis of isatins[13]. This methodology consists in the formation and subsequent oxidation of an intermediate 3-methylthio-2-oxindole to give the corresponding substituted isatins (**4**) in 40-81% yield.

Two complementary methods for the synthesis of the 3-methylthio-2-oxindoles were developed, and the methodology of choice is dependent upon the electronic effect of substituents bonded to the aromatic ring. When electron-withdrawing groups are present, the oxindole derivative can be synthesized via *N*-chloroaniline intermediate, which further reacts with a methyl thioacetate ester to furnish an azasulfonium salt (Method 1, Scheme 4). In the case of electron-donating groups that destabilize the *N*-chloro intermediate, and thus give diminished yields of the azasulfonium salt, a second method of generation of this salt, by reaction of the chlorosulfonium salt with an appropriate aniline, gives better yield of the 3-methylthio-2-oxindoles.

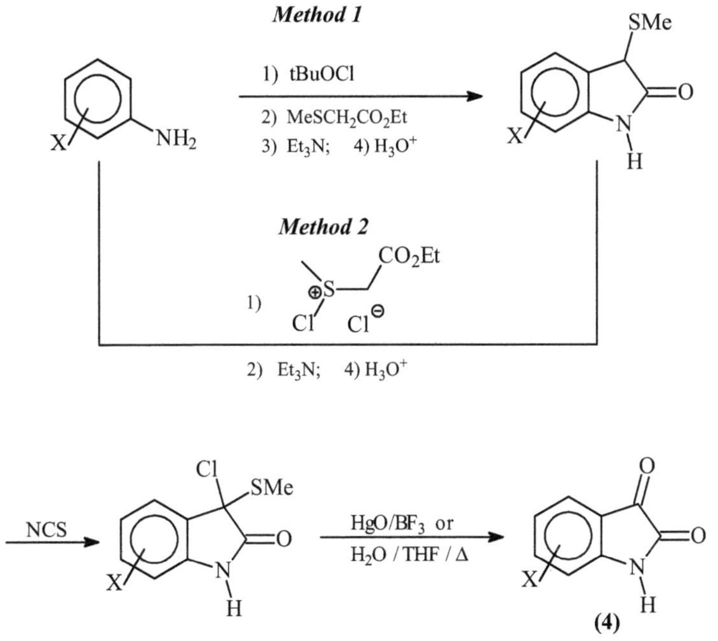

Scheme-4

Reactions of Isatin :

The reactive carbonyl group at position-3 undergoes typical reactions with the ketonic reagents such as hydroxylamine[14], phenyl hydrazine[15] and semicarbazide[16]. The carbonyl group at position-2 is less active and has less ketonic character compared with carbonyl group at C_3.

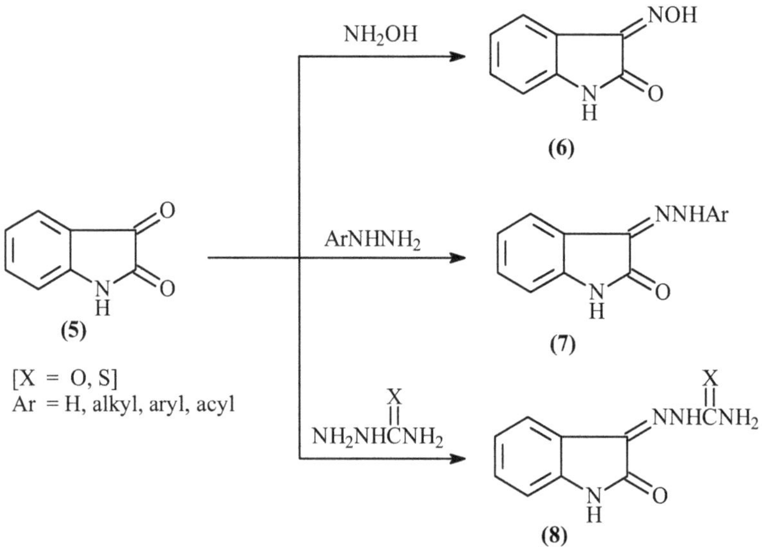

(6)

(5)

[X = O, S]
Ar = H, alkyl, aryl, acyl

(7)

(8)

Schiff Bases :

Isatin reacts with a variety of aromatic amines[17] in the presence of glacial acetic acid to yield Schiff bases **(9)** (azomethines, imines, anils).

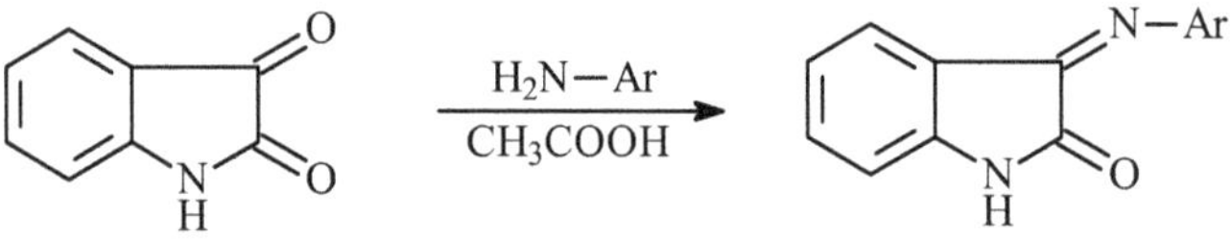

(9)

Isatin-3-imine

Mannich Reaction[18] :

Isatin reacts with formaldehyde and a variety of amines in the Mannich reaction to give their respective Mannich bases **(10)**, in the absence of an amine, isatin and substituted isatin with formaldehyde give hydroxymethyl isatins **(11)**.

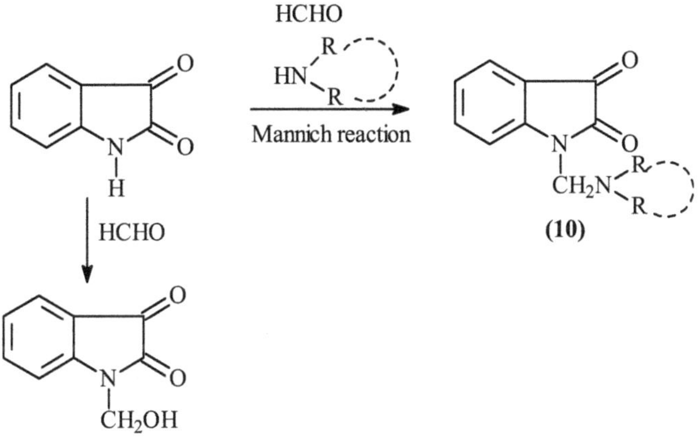

Other Reactions of Isatins
a) Electrophilic Substitution and Related Reactions :

Reaction of Isatin and haloisatins with chlorosulfonic acid gave 5-chlorosulfonylisatin which were then converted into sulfonamides[19].

Nitration of isatin[20], 7-methyl isatin, 4,7-dimethyl isatin[21] and 1-ethyl isatin[22] leads to the introduction of the nitro group into the 5-position. Nitration of 5-methyl isatin gave 5-methyl-7-nitro isatin[24].

Pharmacology of Isatin[25]:

Isatins are endogeneous compounds identified in human and rat tissues. Isatin is a well known pharmacological agent having a range of action in the brain and also protective against certain types of infections.

Isatin contains indole nucleus as that of 5-HT transmitter and is expected to have CNS activity.

Physiologically CNS is divided into following parts[26] :

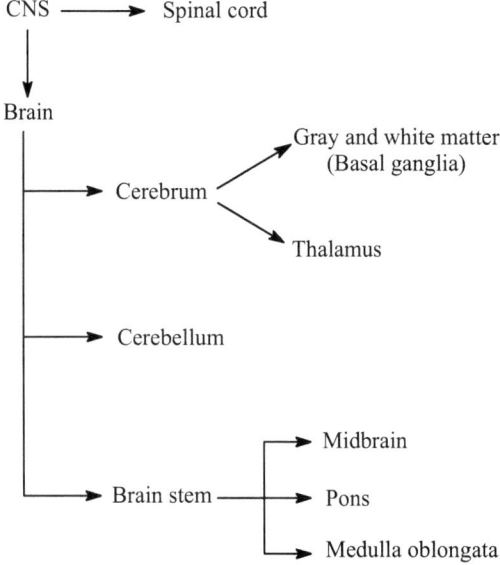

The human brain has about 100 billion nerve cells, also called neurons. Neurons carry signals around the brain and between the brain and the rest of the body. Each neuron produces electrical signals, but chemicals called neurotransmitters are responsible for spreading them[27].

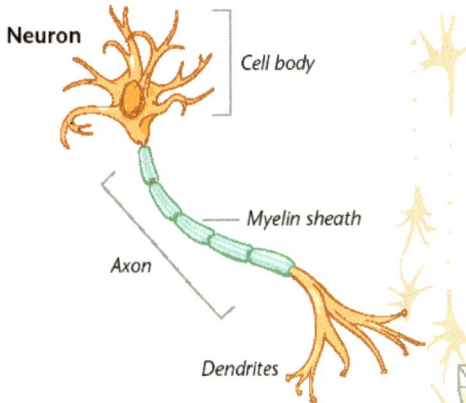

Neuron

Cell body

Myelin sheath

Axon

Dendrites

The electrical signal arises in the neuron's cell body and travels down the axon. At the end of the axon is a space, called the synapse (SIN-apse), between the axon of the first neuron and the dendrites of the next neuron.

At the end of the axon, a neurotransmitter is released. This chemical crosses the synapse and triggers receptors on the dendrites of the next neuron. The next neuron is stimulated to "fire"— producing an electrical signal — and the spread continues.

Not all neurotransmitters are the same. The ones that cause neurons to fire are called excitatory (ek-SI-tuh-TOR-ee) because they excite or increase brain activity. But other neurotransmitters cause neurons to stop firing.

These are called inhibitory (in-HIB-ih-TOR-ee) because they inhibit (block) firing, so there is less electrical activity in the brain.

According to one theory, epilepsy is caused by an imbalance between neurotransmitters that cause neurons to fire and those that cause them to stop firing.

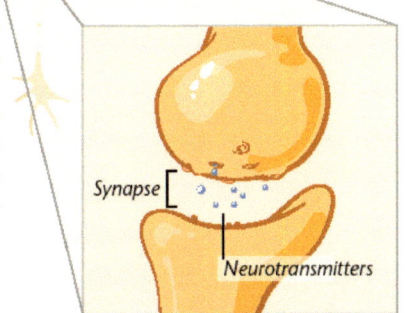

Synapse

Neurotransmitters

The important neurotransmitters of the central nervous system[28]

Name	Predominant activity
γ-Aminobutyric acid (GABA)	Inhibitory
Glycine	Inhibitory
Aspartate	Excitatory
Acetylcholine	Inhibitory
Dopamine	Inhibitory
Norepinephrine	Inhibitory
Epinephrine	Complex
5-Hydroxytryptamine (5-HT)	Inhibitory
Histamine	Complex

The drug acting on CNS are divided into two types[29] :

CNS depressants

- General Anaesthetics
- Anxiolytic, sediative and hypnotic agents
- Antipsychotics
- Anticonvulsant (or) antiepileptic drugs

Central Nervous System Stimulants

- Analeptics
- Methylxanthines
- Central sympathomimetic agents (Psychomotor stimulants)

Mechanism of action of CNS Drugs[30]

• Intereference with the action-potential in the pre-synaptic fibre
  Eg : sedative & hypnotic

• Inteference with ion-channels in pre-or postsynpatic membrane.
  Eg : General anaesthetics

• Interference with the synthesis or storage of neurotransmitter.
  Eg : Reserpine depletes NA

• Intereference with the release or the metabolism of neurotransmitter
  Eg : Antidepressants inhibit the enzyme MAO.

• Blockade of the reuptake of the neurotransmitter. Eg : antidepressants inhibit the reuptake of NA or 5-HT.

• Stimulation or inhibition of specific receptors. Eg : Phenothiazine

EPILEPSY

Epilepsy, characterized by the periodic and unpredictable occurrence of seizures is the most prevalent neurological disorder[31]. Approximately 1% of the world's population has epilepsy, the second most common neurological disorder after stroke. Although standard therapy permits control of seizures in 80% of these patients, millons (500,000 people in the USA alone) has uncontrolled seizures[32].

The term seizure refers to a transient alteration of behaviour due to the disordered, synchronous and rhythmic firing of populations of brain neurons[33]. The term epilepsy refers to a disorder of brain function characterized by the periodic and unpredictable occurrence of seizures.

Seizures can be "nonepileptic" when evoked in a normal brain by treatments such as electroshock or chemical convulsants or "epileptic" when occurring without evident provocation.

Based on the type of behavior and brain activity, seizures are divided into two broad categories, generalized and partial (also called local or focal)..These two categories include many individual types, usually identified by the kind of behavior the seizure produces.

Types of generalized seizures[34]

- Grandmal seizures
- Absence seizures
- Myoclonic seizures
- Clonic seizures
- Tonic seizures
- Atonic seizures

Types of partial seizures

- Simple partial seizures
- Complex partial seizures
- Focal seizures

Generalized seizures are produced by electrical impulses from throughout the entire brain, whereas partial seizures are produced (at least initially) by electrical impulses in a relatively small part of the brain[35].

Generalized seizures (produced by the entire brain)	Symptoms
1.Grandmal or generalized tonic-clonic	Unconsciousness, convulsions, muscle rigidity
2.Absence	Brief loss of consciousness
3.Myoclonic	Sporadic (isolated), jerking movements
4.Clonic	Repetitive, jerking movements
5.Tonic	Muscle stiffness, rigidity
6.Atonic	Loss of muscle tone

Partial Seizures (Produced by a small area of brain)	Symptoms
1.Simple(awareness is retained) a. Simple Motor b. Simple sensory c. Simple psychological	**a.** Jerking, muscle rigidity ,spasms, head-turning b .Unusual sensations affecting either the vision, hearing, smell taste or touch. c. Memory or emotional disturbances.
2.Complex (impairment of Awareness)	Automatisms such as lip smacking, chewing, fidgeting, walking, and other repetitive, involuntary but coordinated movements.
3.Partaial seizure with secondary Generalization	Symptoms that are initially associated with a preservation of consciousness that than evolves into a loss of consciousness and convulsions.

ANTIEPILEPTICS AGENTS[36] .

Pharmacological agents in current clinical use inhibit seizures, and thus referred to as antiseizure drugs . Wheather any of these agents has prophylactic value in preventing development of epilepsy (epileptogenesis) is uncertain.

Accordingly antiepileptic may be classified as follows[37] :

1. Drugs inhibiting Na+ channels :

    (a) Phenytoin, Fosphenytoin, Ethotoin, Mephenytoin

    (b) Lamofrigine, Topiramate

    (c) Carbamezepine

2. Drugs affecting Ca++ channels

    (a) Flunarizine

    (b) Ethosuccinimide, Phensuccinimide, Methsuccinimide

3. Drugs affecting GABA

    (a)Barbiturates like Phenobarbitone, Mephobarbitone, Primidone

    (b) sodium valproate, vigarbetrine (inhibitors of GABA transaminase)

    (c) Tigabine (inhibits GABA reuptake)

    (d) Progabide (GABA receptor stimulator)

    (e) Gabapentene (GABA release)

4. Drugs that bind Benzodiazepine receptor like Clonazepam , diazepam Oxazepam and Midazolam

5. NMDA receptor antagonist: Phencyclidine.

Mechanism of action of Antiepileptic agents[38]

Current antiepileptic drugs are thought to act mainly by two mechanisms:

1. Reducing electrical excitability of cell membranes, possibly through use-dependent block of sodium channels.

2. Enhancing GABA –mediated synaptic inhibition . This may be achieved by an postsynaptic action of GABA-transaminase or by drugs with direct GABA-agonist properties.

3. A few drugs appear to act by a third mechanism, namely inhibition of T type calcium channels.

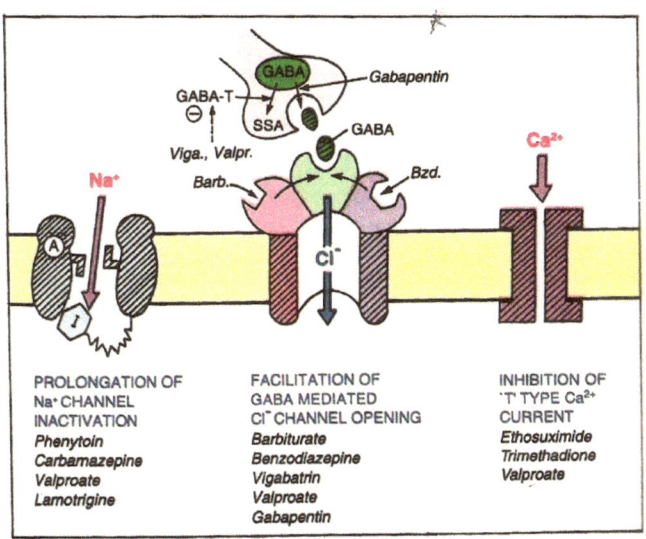

Fig. : Major mechanisms of anticonvulsant action[39]

A : Activation gate; I : Inactivation gate; GABA-T :
GABA transaminase; SSA : Succinic semialdehyde

LITERATURE SURVEY OF BIOLOGICALLY ACTIVE ISATIN DERIVATIVES

ISATIN AS ANTIVIRAL AGENTS

Shin-Hun Juang *et al.*[40] were synthesized N-substituted isatin derivatives (**12**) from the reaction of isatin and various bromides via two steps. Bioactivity assay results (*in vitro* test) demonstrated that some of these compounds are potent and selective inhibitors against SARS coronavirus 3CL protease with IC_{50} values ranging from 0.95 to 17.50 μM

(**12**)

R^1 = H, Cl; R^2 = H, OCH_3, F, I, NH_2
R^3 = Br, NO_2, Cl, NO_2

Luhua Lai and Lu Zhou[41], reported, a series of N-substituted isatin derivatives (**13**) were synthesized and tested against SARS CoV 3C-like protease using a colorimetric assay and confirmed by HPLC. The compounds were shown to be noncovalent reversible inhibitors of SARS CoV 3C-like protease. The C-5 position was found to favour a carboxamide group and N-1 position to favour large hydrophoretic substituents.

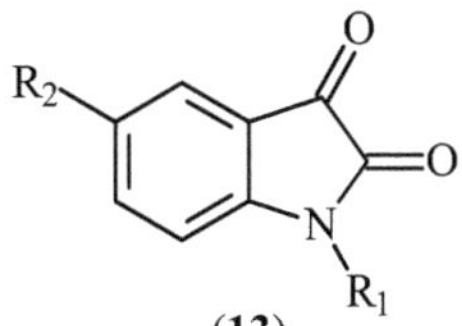

(13)

R_1 = H, CH_3, $CH_3CH_2CH_2$, n-C_4H_9, $PhCH_2$, β-$C_{10}H_7CH_2CH_3CH_2CH_3$
R_2 = I, CO_2CH_3, CO_2H, $CONH_2$

The lowest IC_{50} value (0.37 μM) was observed with compound if and it was selective for SARS CoV 3C-like protease over other protease.

ISATINS AS ANTIMICROBIAL AGENTS

Padhy et al.[42] reported the synthesis of 1-acetyl-3-(2-acetoxy-3-substituted propyloximine)indol-2-(3H)-ones (14) and evaluated for antimicrobial activity against B. subtilis, E. coli and C. albicans. Among them, the compound with 2-acetoxy-3-substituted propyloximino chain at 3-position of indole-2,3-dione exhibited high activity with MIC of 0.35 μg/ml to 12.5 μg/ml against B. subtilis, 0.16 – 3.12 μg/ml against E. coli and 6.2 – 100 μg/ml against C. albicans.

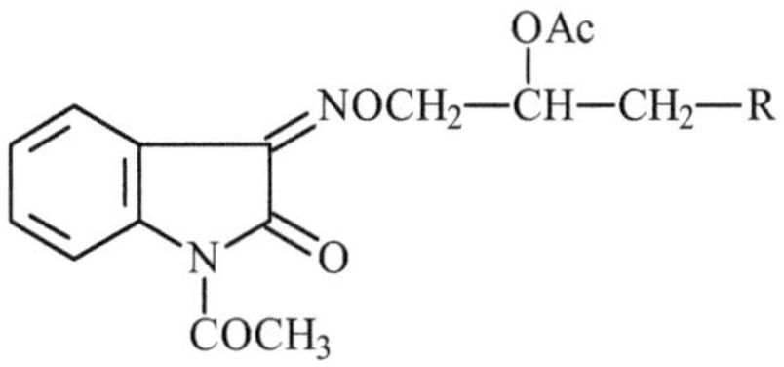

(14)

R – piperidino; pyrrolidino; dicyclohexylamino; diphenylamino;
   methylphenylamino.

Raviraj A. Kusanur *et al.*[43] reported the synthesis of 4'-(coumarin-3-yl)spiro[3H-indol-3,2'-1,5-benzodiazepine]-2(1H)-one (**15**) and evaluated for *in vitro* antibacterial activity.

(15)

R = H, 6-CH$_3$, 8-OCH$_3$, 5,6-benzo, 6-Cl, 6-Br; R$_1$ = R$_2$ = H, CH$_3$

Among all the compounds, compound with R = 8-OCH$_3$, R^1 = R^2 = CH$_3$ showed 88.83% of inhibition against *B. subtilis* and 77.77% of inhibition against *E. coli* as compared to the standard and other compounds were moderately active.

Gupta *et al.*[44] reported the synthesis of 1,3-dihydro-5-substituted-3-[[4-[1-(p-sulphamylphenylimino)ethyl)phenyl]imino]-2H-indol-2-ones (**16**) and screened them for *in vitro* antibacterial activity. Compounds with X = H, R = diazino; X = Cl, R = H; X = Cl, R = -C(=NH)NH$_2$; X = Cl, R = diazino and X = Cl, R = dimidino showed highest activity against *B. subtilis* and *S. aureus*.

(16)

X = -H, Cl, -CH$_3$
R = H, -C(=NH)NH$_2$,

Ajitha *et al.*[45] reported the synthesis of new 2-substituted-[1,3,4]-oxadiazino-[6,5-b]-indoles (**17**). The compounds were evaluated for *in vitro* antimicrobial activity. Compounds (R = Cl, R_1 = H, R_2 = benzimidazolyl), (R = Br, R_1 = H, R_2 = benzimidazolyl) showed highest antimicrobial activity which was comparable with that of Ampicillin.

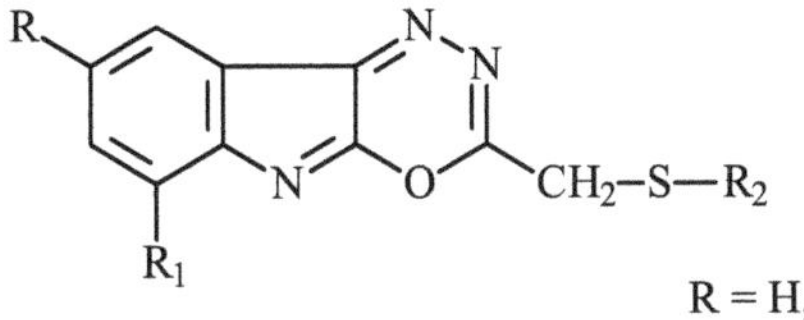

R = H, CH$_3$, Cl, Br
R_1 = H, CH$_3$
(**17**)
R_2 = benzimidazolyl; 4,5-diphenylimidazolyl
5-phenyl-1,3,4-oxadiazolyl
5-phenyl-1,3,4-thiadiazolyl

ISATIN AS ANTICANCER AGENT

Kara L-vine *et al.*[46] reported the *in vitro* cytotoxicity evaluation of some substituted isatin derivatives (**18, 19**).

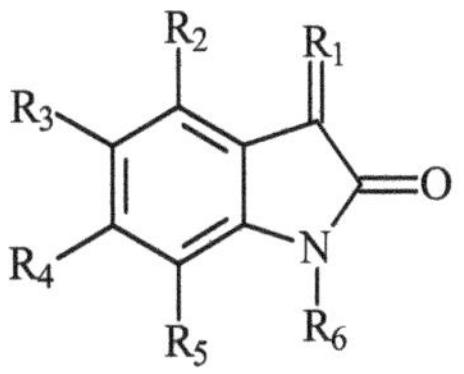

(**18**)

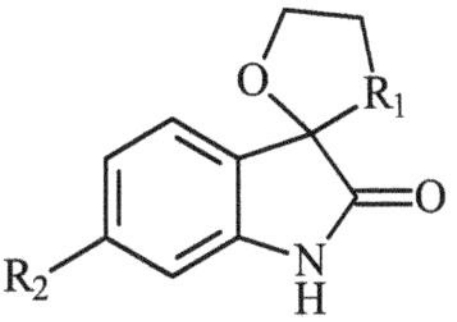

(**19**)

R_1 = O, N-C$_6$H$_5$, N-NHC$_6$H$_5$
R_2 = H, Br; R_3 = H, Br, F, I, NO$_2$, OCH$_3$
R_4 = H, Br, R_5 = H, Br, NO$_2$, R_6 = CH$_3$

Twenty three compounds were synthesized and tested four were selected for further screening against a panel of five human cancer cell lines. These compounds, in general, showed greater selectivity reward leukemia and lymphoma cells over breast, prostate and colorectal carcinoma cells.

Sriram *et al.*[47] reported a series of isatin β-thiosemicarbazone (**20**) derivatives were synthesized a evaluated for their anti-HIV activity in HTLV-III$_B$ strain in the CEM cell line. Three compounds showed significant anti-HIV activity, with an IC$_{50}$ value of 2.62 µM and a selectivity index of 17.41, while net being cytotoxic to the cell line at a IC$_{50}$ value of 44.90.

$$N-NH-\overset{\overset{\displaystyle S}{\|}}{C}-N\overset{C_2H_5}{\underset{C_2H_5}{}}$$

$$=O$$

$$N$$
$$-R^1$$

$$R^1 = \quad -N\overset{CH_2-C_6H_5}{\underset{CH_2-C_6H_5}{}} \qquad -N\overset{CH_3}{\underset{CH_3}{}} \qquad -N\overbrace{\qquad}N-CH_2-C_6H_5$$

$$-N\overbrace{\qquad}N-\overbrace{\qquad}-F$$

(20)

Klaus Kopka *et al.*[48] reported 5-pyrrolidinyl sulfonyl (**21**) isatins as a potential tool for the molecular imaging of carpases in apoptosis.

R = CH$_3$, Ph

(21)

ISATIN AS CNS AGENT

CNS depressant activities. – Isatin has a range of actions such as CNS-MAO inhibition, anticonvulsant and anxiogenic activities. Its effect as a mono amino-oxidase (MAO) inhibitor is the most potent *in vitro* action recorded to date. It is a selective MAO B inhibitor with an inhibitory concentration (IC_{50}) of about 3 µg mL^{-1}[(49)]. At higher concentrations it inhibits a variety of other enzymes, such as alkaline phosphatases. In rodents, it has been reported to act as an antiseizure agent and to potentiate the antiseizure action of propranolol. Isatin has also been found to increase vigilence[50]. At a low dose (15 mg kg^{-1}), it is anxiogenic and prolongs pentylenetetrazole (PTZ) induced seizures while at higher dosage (80 mg kg^{-1}) it becomes sedative and anticonvulsant and the brain 5-HT levels are found to be significantly raised[51].

Anticonvulsant activity. – Bhattacharya and Chakraborti[52] reported isatin to be an endogenous compound with anxiogenic properties, which occur within a narrow intraperitoneal (*i.p.*) dose range (15–20 mg kg–1). Higher doses exhibited a significant anticonvulsant effect against both PTZ and 3 MPA (mercapto propionic acid) induced clonic convulsions. Bhattacharya *et al.*[53] have found isatin to function as a potent antagonist on anti-natriuretic peptide (ANP) receptors *in vitro,* and to inhibit anxiolytic, memory facilitating and diuretic actions of intracerebroventricularly administered ANP.

21

Olesen and Kanstrup[54] prepared pyrido[2,3-b]indoles to treat a disease in the CNS *via* the metabotrophic glutamate receptor system. The title compounds are useful for treating diseases in the CNS such as epilepsy, senile dementia and Parkinsonism (22).

R^1 = H, C_{1-6} alkyl (CH$_3$, C_2H_5, C_3H_7-C_6H_{13}); C_{2-6} alkenyl (C_2H_4, C_3H_6, -C_6H_{12}), R^2 = piperidino, morpholino; R^3 = H, COOH, CN; R^4 = H, C_{1-6} alkyl (CH$_3$, C_2H_5, C_3H_7); R^5 = R^8 = H, NO$_2$, NH$_2$

(22)

Evanno *et al.*[55] synthesized 1*H*-pyrido[3,4-b]indole-4-carboxamide derivatives of the structure presented in (23).

X = H, holo, axlkyl, alkoxy, CF$_3$, OCH$_3$; R^1 = H, alkyl, cyclopropyl, CH$_3$; R^2 = alkyl, phenyl alkyl, cyclo-hexylmethyl, thienylmethy; R^3 = R^4 = H, alkyl, 2-methoxy ethyl, OC$_2$H$_5$, carboxy alkyl, alkoxycarbonylakyl, phenyl alkyl, pyrrolidinyl, piperidinyl, morpholinyl, 4-methyl piperazinyl azetidinyl, thiadiazolinyl

(23)

The different substituted compounds (23) were tested for their anxiolytic, hypnotic and anticonvulsant activities.

Evanno et al.[56] synthesized 4-oxo-3,5–dihydro-4-H-pyridazano-4,5-b-indole-1-acetamide derivatives that can be used for treating diseases related to GABA aminergic transmission disorders. The compounds also shows hypnotic and anticonvulsant activities in rats and mice. The structures of the compounds are given in (24).

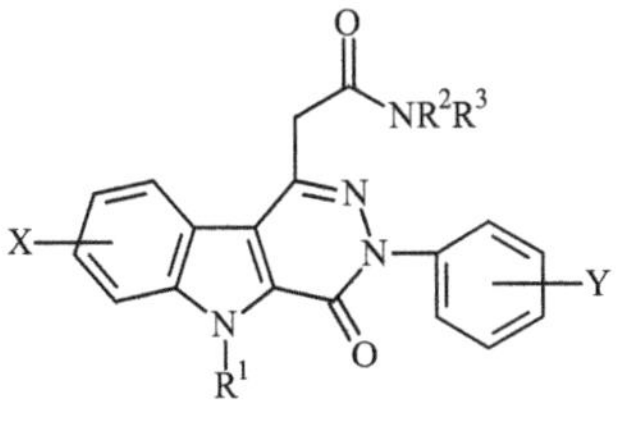

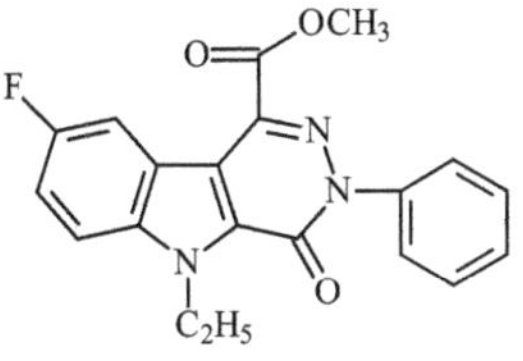

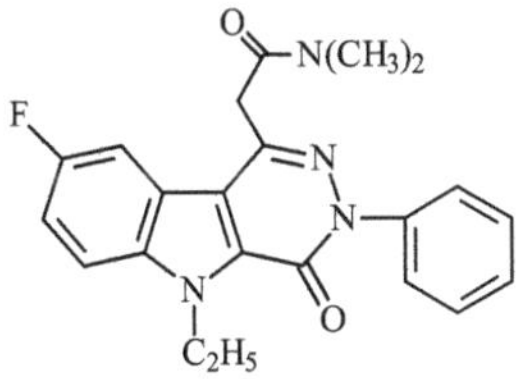

X = H, halo, CH$_3$, OCH$_3$, OCH$_2$C$_6$H$_5$; Y = H, 1 halo atom, CH$_3$, OH, OCH$_3$, NO$_2$; R^1 = H, C$_{1-4}$ alkyl (CH$_3$, C$_2$H$_5$, C$_3$H$_7$, etc.); R^{2}R^3 = H, C$_{1-4}$ alkyl, CH$_2$C$_6$H$_5$ or R^{2}R^3 = azetidine, pyrrolidinyl, 3-ethoxy pyrrolidinyl, piperidinyl, morpholinyl, 4-methyl piperazinyl, 1,3-thiadiazolinyl

(24)

David et al.[57] have shown that 2-aminonapthyridine is prepared by ring cleavage of 2-isoindolinyl napthyridine (25).

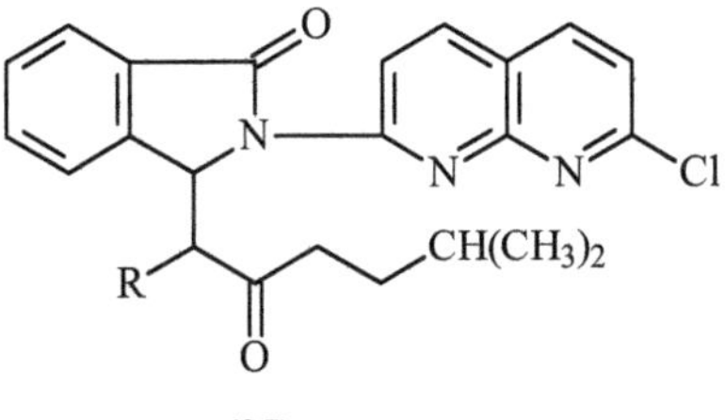

R = H, COOH

(25)

These compounds have exhibited remarkable anxiolytic, hypnotic, anticonvulsant and muscle relaxant properties.

Srivastava *et al.*[58] synthesized a series of compounds from carbazole, which on condensation with chloroacetyl chloride in the presence of triethylamine afforded azetidinones. Some of the compounds exhibited promising antibacterial, antifungal, anti-inflammatory and anticonvulsant activities (**26**).

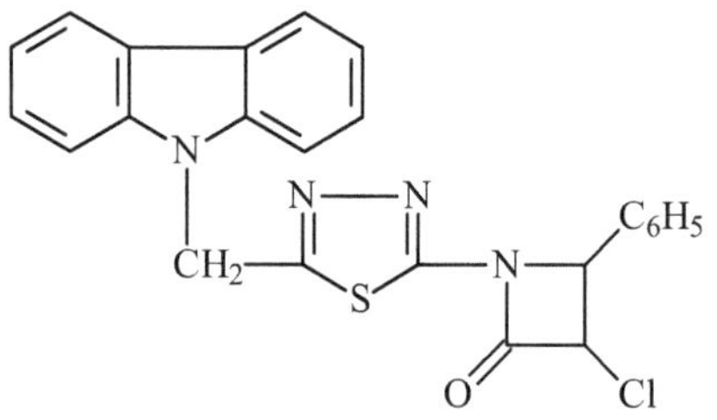

(**26**)

Singh *et al.*[59] synthesized a series of isatin-based spiroazetidinones and screened them for their anticonvulsant activity (**27**).

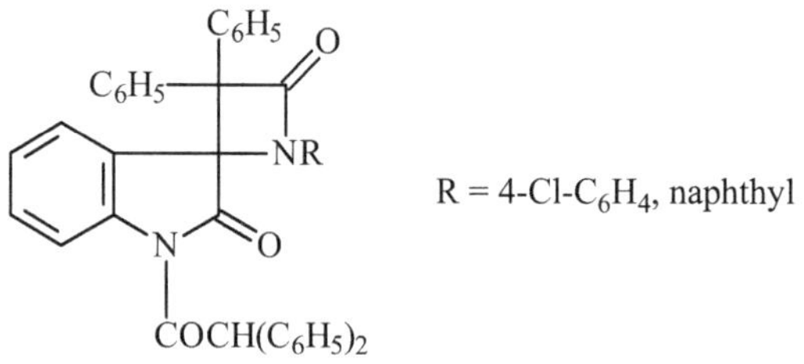

R = 4-Cl-C_6H_4, naphthyl

(**27**)

Pandeya *et al.*[60] synthesized a series of *p*-nitrophenyl substituted semicarbazones and their anticonvulsant activity was screened against maximal electroshock (MES), subcutaneous pentylenetetrazole (ScPTZ) and subcutaneous strychnine (ScSTY) tests (**28**).

All the compounds were active in subcutaneous pentylenetetrazole and MES tests. Two compounds were active in the MES test at 100 mg kg–1.

Pandeya *et al.*[61] synthesized a series of *N*-methyl/acetyl-5-(un)substituted-isatin-3-semicarbazones (**28**).

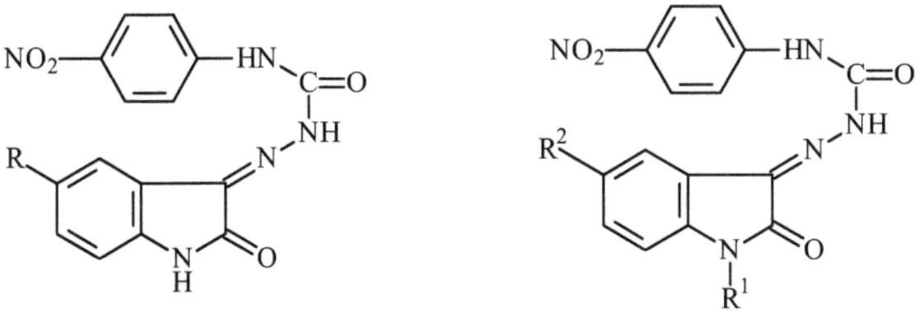

R = H, Cl, NO$_2$ R = 2-Cl, 3-Cl, 4-Cl, 4-Br, 4-NO$_2$, 4-SO$_2$NH$_2$
R^1 = CH$_3$, COCH$_3$, R^2 = H, NO$_2$

(28)

In this series, compounds with 4-bromo and 2-chloro substitution (R = 4-Br and 2-Cl) showed promising activity and were also active in MES, subcutaneous pentylenetetrazole and subcutaneous strychnine induced tests.

Further, Pandeya *et al.*[62] synthesized halosubstituted isatin semicarbazones to study the role of hydrogen bonding for anticonvulsant activity (**29**).

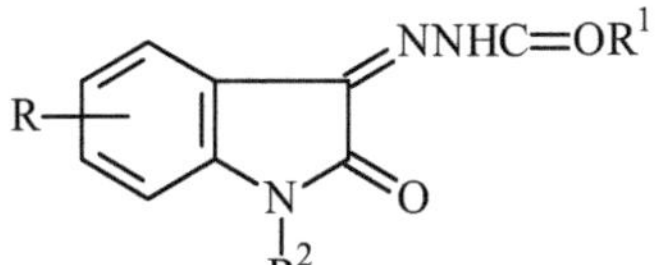

(29)

R^1 = -CH$_2$O-C$_6$H$_5$, 1-CH$_2$O-(4-Br)-C$_6$H$_4$,
R^2 = COCH$_3$, CH$_2$C$_6$H$_5$, R = H, 4-Cl, 5-Cl, 6-Cl

Pandeya et al.[63] had synthesized Schiff bases of *N*-methyl and *N*-acetyl isatin derivatives with different aryl amines and screened them for anticonvulsant activity against MES and scMet. *N*-methyl-5-bromo-3-(*p*-chlorophenylimino) isatin exhibited anticonvulsant activity in MES and scMet with *LD50* > 600 mg kg–1, showing better activity than the standard drugs such as phenytoin, carbamazepine and valproic acid (**30**).

R = Br, NO_2 ; R^1 = CH_3, $COCH_3$;
R^2 = NO_2, COOH, OCH_3, Cl, F

(**30**)

Anxiogenic and other CNS activities. – Palit *et al.*[64] studied the behavioural effects of isatin, a putative biological factor in rhesus monkeys. Isatin, one of the constituents of tribulin, a postulated endocoid marker of stress and anxiety, has been shown to induce anxiety in rodents. Medvedev *et al.*[65] studied a range of isatin analogues for their *in vitro* inhibition of human MAO A and B. Most analogues were less potent than isatin. Hydroxylation of the aromatic ring in isatin changed the inhibitory potency in favour of MAO A, with 5-hydroxy isatin being a potent and selective MAO A inhibitor (IC_{50} 8 µg mL^{-1}). Isatinic acid, which is formed reversibly from isatin in alkaline medium, showed no inhibition (**31**).

(**31**)

Kennis et al.[66] synthesized hexahydropyrido (4,3-b) indole derivatives.

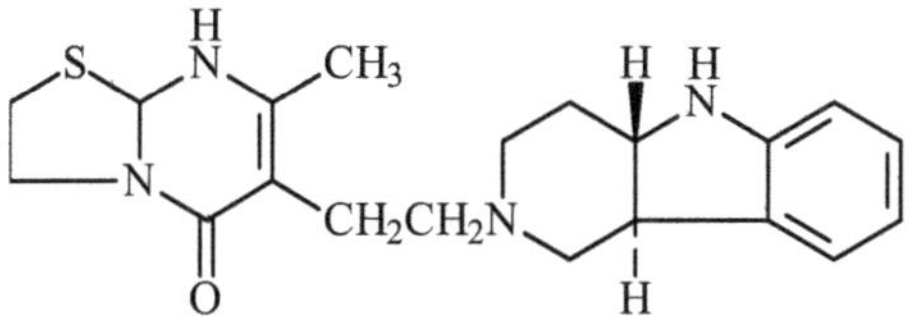

(32)

The compound displayed in (**32**). was found to have central dopamine and serotonin antagonistic activity in the combined apomorphine, tryptamine and nor-epinephrine test in rats.

Sarangapani et al.[67] reported the synthesis of 3-methyl-4-(oxindol-3-ylidenyl)-5-pyrazolones (**33**). All compounds screened for gross behaviour studies they exhibited CNS depression, reduced locomotor activity. Compounds with 5-CH_3 group on indolinone and 3-methyl group on pyrazole showed antibacterial activity.

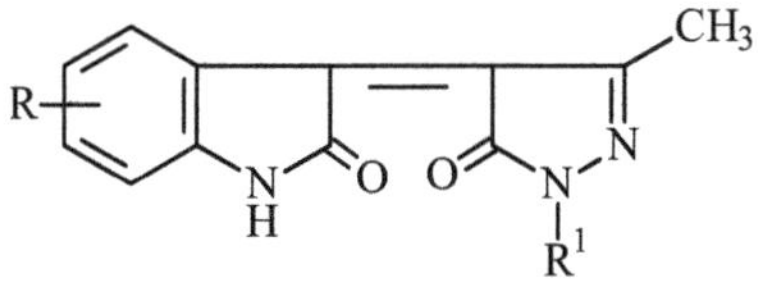

(R = H, 5-CH_3, 5-Cl; R^1 = H, C_6H_5)

(33)

Maria Eline Mathew[68] has designed to investigate the inhibitory effect of isatin derivatives on lipopolysaccharide / intereferon-γ-induced expression of inducible nitric oxide synthase (iNOS) and cyclooxygenase-2- (Cox-2) proteins, production of prostaglandin E_2 (PGE$_2$), nitric oxide (NO), tumor necrosis factor (TNF-α) and their capacity to scavenge No. Isatins (**34**) inhibit TNF-α production and iNOS and COX-2 protein expression resulting on reduced levels of No & PGE$_2$.

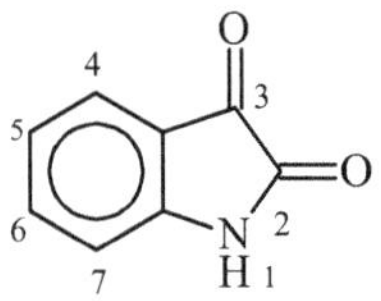

(R = H, 5-F, 5-Cl, 6-Cl, 7-Cl, 4-Br, 5-I, 5-CH$_3$)

(34)

Giscle Zapata *et al.*[69] reported the synthesis of novel isatin ketals (nine dioxolane ketals and nine dioxane ketals) (**35**) and studied for their sedative, hypnotic and anesthetic effects using pentobarbital induced sleeping time, locomotor activity.

They observed dioxolane ketals were more potent than dioxane ketals for inducing sedative-hypnotic states, causing upto a three-fold increase in pentobarbital hyperosis.

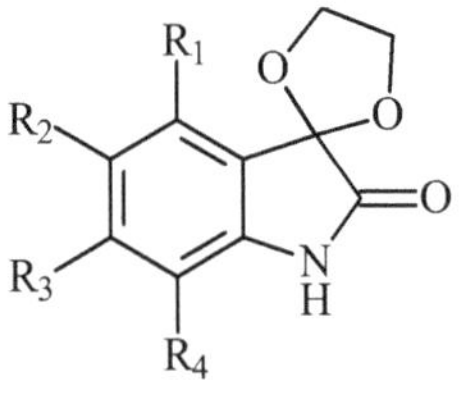

(35)

R$_1$ = H, Cl, Br; R$_2$ = H, Cl, Br, F
R$_3$ = H, Br ; R$_4$ = H, Cl, CF$_3$, OCH$_3$, Br

Sasmal *et al.*[70] reported synthesis and evaluation of CNS activity of some spiro isationoid compounds (**36**). All compounds exhibited CNS depression activity.

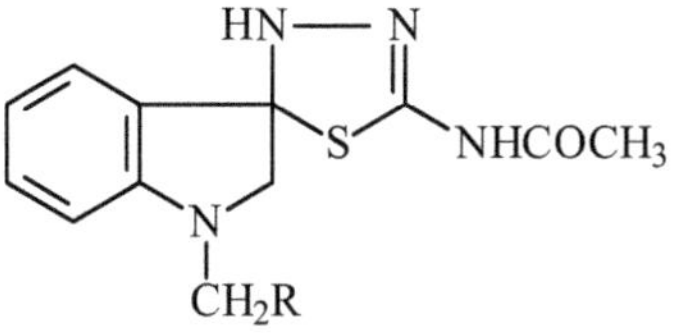

(36)

R = -N(CH$_3$)$_2$, -N(C$_2$H$_5$)$_2$, -N(morpholine), -N(piperidine), —N (C$_6$H$_5$)$_2$

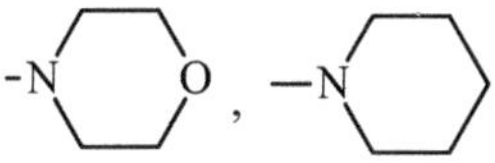

Isatins as Psychotropic Agents

Varma *et al.*[71] reported the synthesis of alkyl-4-[4'-(1,2-dihydro-5-chloro-2-oxo-3H-indol-3-ylidene amino)benzoyl]aminobenzoate (**37**) and the Mannich bases of these compounds (**38**) with morpholine were found to be non-toxic and psychotropic. Some of these compounds were also active against mycobacterium tuberculosis.

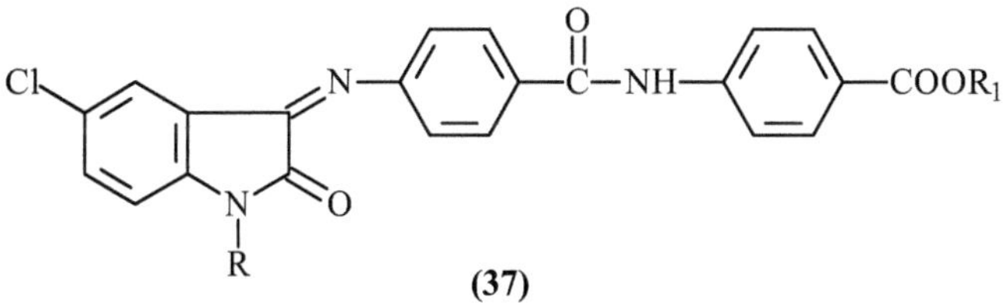

(37)

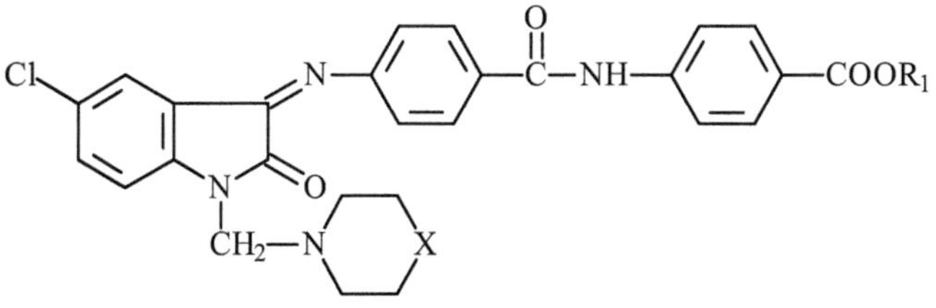

(38)

Effect of indoles on pentobarbitone induced sleeping time

Sarangapani *et al.*[72] reported eight new 2-substituted-[1,3,4]oxadiazino-[6,5-b]indoles (**39**). All the compounds exhibited reduction in locomotor activity and potentiation of pentobarbitone sodium induced sleeping time in experimental animals.

(39)

R = Cl, Br R_1 = CH$_3$ R_2 = H, CH$_3$

CHAPTER-2

AIMS AND OBJECTIVES OF THE PRESENT WORK

It is evident from literature that the presence of the indole nucleus found to have various pharmacological activities like antimicrobial, anti-convulsant, MAO inhibitory, anticancer and psychotropic activities. It is interesting to note from the literature that the important isatin moiety is yet to be explored both synthetically and biologically in conjunction with the molecular moieties of several other biologically prominent heterocyclic systems, inspite of the extensive work reported so far on isatins.

Therefore, keeping this in view as the main objective, the present project has been aimed at achieving the following :

➢ To synthesize the Isatin derivatives as depicted in **Scheme**-I.
➢ To purify the new isatin derivatives by crystallization and chromatographic technique.
➢ To characterize the new compounds by physical and spectral data (IR, ^{1}H NMR and Mass)
➢ To screen the compounds for CNS activity by standard protocol available in literature
➢ To study the SAR of the compounds.

PLAN OF WORK

In the present investigation, involving reactions of isatin with a view to synthesize some biologically active compounds, it has been felt worthwhile to study chlorosulphonation of isatin to form Isatin-5-sulphonic acid chloride and it is subjected to reaction with different amines or anilines to form respective sulphonamide derivatives, as such a reaction is not reported so far and also to evaluate the products biologically. The synthesis of title compounds could be achieved by the Scheme – I.

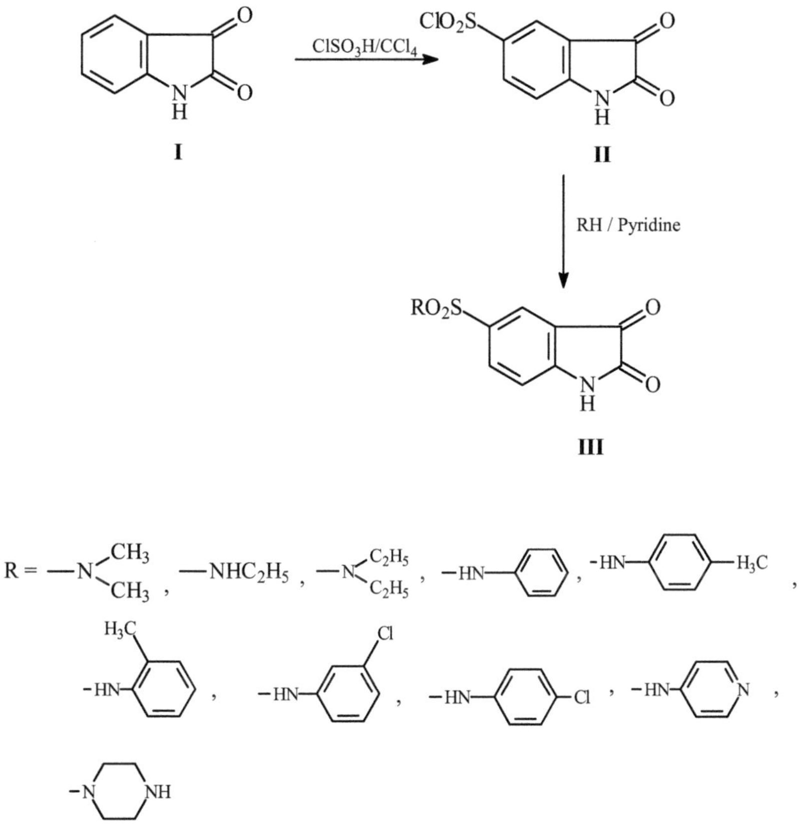

$$\text{I} \xrightarrow{\text{ClSO}_3\text{H/CCl}_4} \text{II} \xrightarrow{\text{RH / Pyridine}} \text{III}$$

I

II

III

$R = $ —N(CH$_3$)CH$_3$, —NHC$_2$H$_5$, —N(C$_2$H$_5$)C$_2$H$_5$, —HN—C$_6$H$_5$, —HN—C$_6$H$_4$—H$_3$C ,

—HN—(H$_3$C)C$_6$H$_4$, —HN—(Cl)C$_6$H$_4$, —HN—C$_6$H$_4$—Cl , —HN—C$_5$H$_4$N ,

—N(piperazine)NH

Scheme – I

CHAPTER-3

EXPERIMENTAL

I. PREPARATION OF ISATIN-5-SULFONIC ACID CHLORIDE (II)

To Isatin (5.9gms; 0.04 moles) cooled in ice, was added chlorosulphonic acid (46.6gms; 0.4 moles) and carbon tetrachloride (5ml; 0.06 moles). The reaction mixture was gradually heated to 70°C and held that temperature for two hours. The deep brown reaction solution was cooled and poured into ice (200gms) to obtain Isatin-5-sulfonic acid chloride(II, 6gms) as a yellow powder. It was crystallized from acetone-benzene. (m.p=145-147^{0}c, yield=80%)

II. PREPARATION OF ISATIN-5-SULPHONAMIDE DERIVATIVES(III):

Reflux a mixture of the amine or aniline, (0.05 mole) and Isatin-5-sulphonic acid chloride **II** (0.05 mole) in 6ml of pyridine for 30 minutes. Cool and pour the reaction mixture into 10 ml of cold water and stir until the product was separated. Filter off the solid and recrystallize it from ethanol.

Adopting this procedure as many as few compounds were prepared and their physical data is given Table-I

Table -I

Physical data of Isatin-5-Sulphonamide derivatives (III)

S.No.	Compound	R	Mol. Formula	Mol. Wt	m.p. (^{o}C)	% Yield
1	IIIa	$-N(CH_3)_2$	$C_{10}H_{10}N_2O_4S$	254.2	170-72	82
2	IIIb	$-NHC_2H_5$	$C_{10}H_{10}N_2O_4S$	254.2	165-67	81
3	IIIc	$-N(C_2H_5)_2$	$C_{12}H_{14}N_2O_4S$	282.3	163-65	78
4	IIId	C6H5—NH—	$C_{14}H_{10}N_2O_4S$	302.3	175-77	85
5	IIIe	(2-CH3)C6H4—NH—	$C_{15}H_{12}N_2O_4S$	316.3	173-75	75
6	IIIf	H3C—C6H4—NH—	$C_{15}H_{12}N_2O_4S$	316.3	175-78	68
7	IIIg	(3-Cl)C6H4—NH—	$C_{14}H_9ClN_2O_4S$	336.74	180-82	72
8	IIIh	Cl—C6H4—NH—	$C_{14}H_9ClN_2O_4S$	336.74	181-83	75
9	IIIi	Pyridyl—NH—	$C_{13}H_9N_3O_4S$	303.1	179-80	81
10	IIIj	HN-piperazinyl—	$C_{12}H_{13}N_3O_4S$	295	163-65	81

SPECTRAL DATA

1. Isatin-5-sulphonic acid chloride (III)

IR (KBr) cm^{-1}: 3445 (NH), 1623 (C = O), 1423 (SO$_2$), 1497 (C-H Bending), 1244.32 (C-O), 830-500 (Ar).

^{1}H NMR (300 MHz, DMSO): δ [ppm]: 6.9 (s, 1H, Ar-H), 7.4 (d, 1H, Ar-H), 7.6 (s, 1H, Ar-H) and 10.5 (s, 1H, N-H).

2. 2,3 dioxo-N-Phenyl-2,3-dihydro-1H-Indole-5-sulphonamide (IIId)

IR (KBr) cm^{-1}: 3364(NH), 3188(NH), 1763 (C=O), 1616(C=C), 1371(SO$_2$).
^{1}H NMR (300 MHz, DMSO): δ [ppm], 6.95(1H, NH, SO2NH), 7.7-8.2 (m, 8H, Ar-H), 11.2 (1H, NH,Lactam)

3. N-(4-methyl Phenyl)-2,3-dioxo-2,3 dihydro-1H-Indole-5-Sulphonamide (IIIf)

IR (KBr) cm^{-1}: 3357 (NH), 2921 (C-H streching), 1727 (C=O), 1330 (SO$_2$).

CHAPTER-4

INTRODUCTION

It is evident from the literature that Isatins exhibited Anticonvulsant, psychotropic, anxiogenic, properties. Therefore, it has been felt worthwhile to screen the new isatin-5-sulphonamide derivatives for CNS activity.

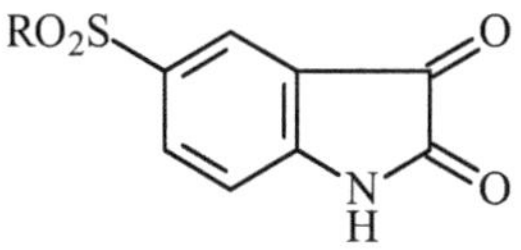

III

For this purpose ten new isatin-5-sulphonamide derivatives (III) were synthesized as described in Scheme-I.

Action on Central Nervous System - Gross Behavioral Studies[73]
Materials
0.1% sodium CMC, test compounds and sonicator.
Animals : Swiss mice

Method :

Healthy albino swiss mice weighing between 20 to 25 gms were used in this investigation. Animals were fasted over night and divided into groups of six animals each. The test compounds suspended in 0.1% sodium CMC were administered by i.p. in doses upto 100 mg/kg body weight. The control group animals received only vehicle (0.1% sodium CMC).

All the test compounds were observed for gross behavioural changes continuously for 7 hrs starting from the administration of test compound and 48 hrs intermittently and compared with that of control group of mice. The gross behavioural studies reveal that all the compound have produced CNS depression in the test animals and no mortality was observed. The data on gross behavioural changes is presented in Table – II.

LOCOMOTOR ACTIVITY[74]

0.1% sodium CMC, test compounds, sonicator and actophotometer.
Animals : Swiss mice.

Method :

The locomotor activity was studied by using actophotometer, which operates on photoelectric cells, which are connected in circuit with a counter. When beam of light falling on the photocells is cut of by animals, a count was recorded. Healthy male mice weighing between 20-25 gm were used. Animals were faster for over night and divided into groups of six animals each. The test compounds suspended in 0.1% sodium CMC were administered at a dose of 100 mg/kg body weight, i.p. The control group animals received only vehicle (0.1% sodium CMC). The response (counts) was recorded after 30 minutes administration of test compound. The animals were placed in actophotometer for 10 minutes and scores were recorded (number of deflections). The results were compared with the control and presented in Table-III .

ANTICONVULSANT ACTIVITY[75]

Maximal Electroshock – Induced Convulsions

Materials :

0.1% sodium CMC, test compounds, sonicator, electroconvulsometer, corneal electrode, stop watch, phenytoin.

Animals : Swiss mice.

Method :

The anticonvulsant activity was studied by using electro-convulsometer. Healthy male mice weighing between 20-25 gm, were fasted for overnight and divided into groups of six animals each. The test compounds suspended in 0.1% sodium CMC were administered at a dose of 100 mg/kg body weight i.p. The control group animals received only vehicle (0.1% sodium CMC). The test started 30 min after i.p. injection. Maximal seizures were induced by the application of electrical current to the brain via corneal electrodes. The stimulus parameter for mice was 50 mA in a pulse of 60 Hz for 200 ms. Abolition of the hind limb tonic extensor spasm was recorded as a measure of anticonvulsant activity, results are presented in Table-IV.

TABLE-II

GROSS BEHAVIOURAL STUDIES

Compounds	Time in hrs	Alertness	Visual Placing	Awareness				Mood		
				Sterotype	Passivity	Writhing	Grooming	Vocalization	Restlessness	Irritability
IIIa	1/2	+	-	-	-	-	-	-	-	-
	1	+	-	-	-	-	-	-	-	-
	2	+	-	-	-	-	-	-	-	-
	3	-	-	-	-	-	+	-	-	-
	4	-	-	-	-	-	+	-	-	-
	5	+	-	-	-	-	-	-	-	-
	24	+	-	-	-	-	-	-	-	-
IIIb	1/2	+	-	-	-	-	-	-	-	-
	1	+	-	-	-	-	-	-	-	-
	2	+	-	-	-	-	-	-	-	-
	3	-	-	+	-	-	+	-	-	-
	4	-	-	-	-	-	+	-	-	-
	5	+	-	-	-	-	-	-	-	-
	24	+	-	-	-	-	-	-	-	-
IIIc	1/2	+	-	-	-	-	-	-	-	-
	1	+	-	-	-	-	-	-	+	-
	2	+	-	-	-	-	-	-	-	-
	3	-	-	-	-	-	+	-	-	-
	4	-	-	-	-	-	+	-	-	-
	5	+	-	-	-	-	-	-	-	-
	24	+	-	-	-	-	-	-	-	-

40

TABLE-II

GROSS BEHAVIOURAL STUDIES(continued)

IIId	1/2	+	-	-	-	-	-	-	-	-	-	-	-	-
	1	+	-	-	-	-	-	-	-	-	-	-	+	-
	2	+	-	-	-	-	-	-	-	-	+	-	-	-
	3	-	-	-	-	-	+	-	-	-	-	-	-	-
	4	+	-	-	-	-	+	-	-	-	-	-	-	-
	5	+	-	-	-	-	-	-	-	-	-	-	-	+
	24	+	-	-	-	-	-	-	-	-	-	-	-	-
IIIf	1/2	+	-	-	-	-	-	-	-	-	-	-	-	-
	1	+	-	-	-	-	-	-	-	-	+	-	+	-
	2	+	-	-	-	-	+	-	-	-	-	-	-	-
	3	-	-	-	-	-	+	-	-	-	-	-	-	-
	4	+	-	-	-	-	+	-	-	-	-	-	-	-
	5	+	-	-	-	-	-	-	-	-	-	-	-	+
	24	+	-	-	-	-	-	-	-	-	-	-	-	-

**Effect of Isatin-5-sulphonamide derivatives(III) on Locomotor Activity
Readings of Actophotometer**

Compounds	Before admn of compound (10min)	After admn of compound (10min)	% change in activity
IIIa	442.63	210.55	52.43
IIIb	462.00	185.12	59.93
IIIc	447.12	161.92	63.78
IIId	470.32	147.65	68.60
IIIf	455.65	127.29	72.00

n = six in each group

All the test compounds were administered in a dose of 100 mg/kg (b.w) i.p. and motor activity in an open field was determined. Data are expressed as the number of movements per minute (Mean) and calculated per cent decrease in activity.

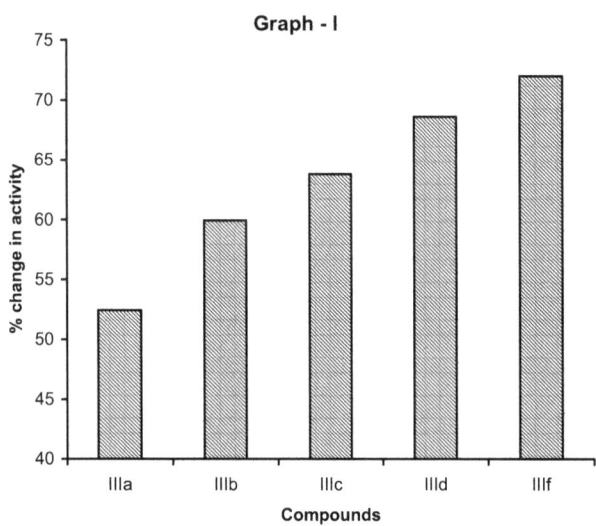

TABLE – IV

Data on anticonvulsant activity of test compounds using Maximal Electroshock (MES) Induced Convulsions

Compound[a]	Duration of tonic hind limb extensor phase (in secs)	Animals protected in %
Control[b]	11.23	0
IIIa	5.6	50.13
IIIb	5.0	55.47
IIIc	4.5	59.90
IIId	4.0	64.38
IIIf	3.8	66.16
Phenytoin[c]	0	100

All animals are recovered

[a] The compounds were tested at a dose of 100mg/kg(i.p.)
[b] 0.1% Sodium CMC at a dose of 1ml/kg (i.p.)
[c] Tested at a dose of 40mg/kg (i.p.)
n = six in each group.

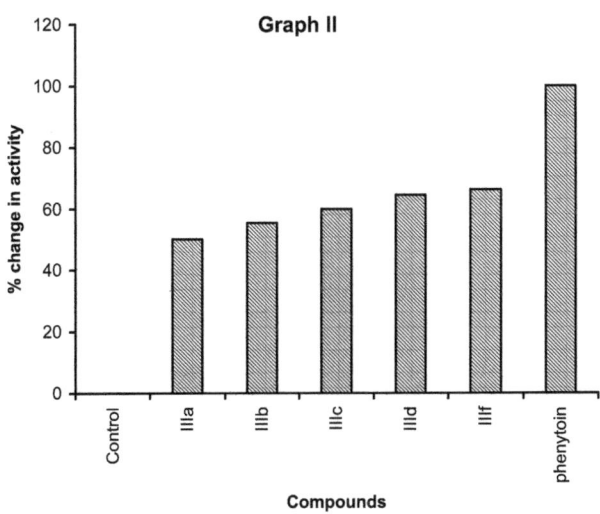

CHAPTER-5

RESULTS AND DISCUSSION

Gross Behavioural Studies :

The gross behavioural studies of the test compounds reveal that all the compounds exhibited CNS depression in the mice.

Effect on Locomotor Activity :

Table II and Graph-I pertaining to the results of the effect of the test compounds on locomotor activity, reveals that compound IIIf (R = p-toluidino) showed more reduction in the locomotor activity which was followed by compounds IIId, IIIc, IIIb, IIIa respectively.

Effect on Anticonvulsant Activity

The data on effect of the test compounds on anticonvulsant activity is presented in Table III. Among the compounds subjected to anticonvulsant activity (Graph-II), it is observed that compounds IIIf exhibited more promising activity,. compounds IIId, IIIc, IIIb, IIIa were found to be next in the order of reducing the duration of convulsions. Compounds with substitutions by diverse electron rich groups on aryl ring increased potency in the Maximal Electroshock Seizure (MES) screening.

CONCLUSIONS

The following conclusions could be drawn broadly from the results of these investigations.

1) Synthetic work could be positive as per the planning and as such in all the reactions carried out, the expected compounds alone could be obtained.

2) All the test compounds showed CNS depression while studying the gross behavioural changes.

3) All the test compounds exhibited reduction in locomotor activity.

4) Compound IIIf (R = p-toluidino) showed more reduction in the locomotor activity among all the test compounds.

5) Compounds IIId, IIIc, IIIb, IIIa were next in the order of reduction of locomotor activity.

6) All the test compounds exhibited anticonvulsant activity. Compound IIIf (R = p-toluidino),showed more promising anticonvulsant activity followed by compounds IIId (R = anilino), IIIc (R = diethylamino), IIIb (R = ethylamino), IIIa (R = dimethylamino).

7) The link between the functions of noradrenaline and 5-hydroxy tryptamine (5-HT) in certain parts of the brain develop the symptoms of depression. Thus it may be concluded that as the test compounds containing indole nucleus and substitution at C-5 position of the Indole nucleus resembles the structureof 5-HT, expected to contribute to the CNS activity.

REFERENCES

1. Erdmann, *J. Prakt. Chem.,* **24** (1841) 1.

2. Laurent, *J. Prakt. Chem.,* **25** (1842) 430.

3. R.J. Sundberg "The Chemistry of Indoles", Academic, New York, 1975.

4. A.V.N. Chenko, A.G. Drushlyak and V.V. Tatov, *Chem. Heterocycl. Comp.,* **10** (1984) 1155.

5. M. Alam, M. Younas, M.A. Zafar and Naeem, *Pak. J. Sci. Indian Res.,* **32** (1989) 246 (CA 112: 7313u).

6. K. Lackey and D.D. Sternbach, *Synthesis,* (1993) 993.

7. K. Lackey, J.M. Besterman, W. Fletcher, P. Leitner, B. Morton and D.D. Sternbach, *J. Med. Chem.,* **38** (1995) 906.

8. W.M. Bryant III, G.F. Huhn, J.H. Jensen and M.E. Pierce, *Synth. Commun.,* **23** (1993) 1617.

9. W.A. Lopes, G.A. Silva, L.C. Sequeira, A.L. Pereira and A.C. Pinto, *J. Braz. Chem. Soc.,* **4** (1993) 34.

10. W.J. Welstead Jr., H.W. Moran, H.F. Stauffer, L.B. Turnbull and L.F. Sancilio, *J. Med. Chem.,* **22** (1979) 1074.

11. A. Taylor, *J. Chem. Res.(S),* (1980) 347

12. K.C. Rice, B.J. Boone, A.B. Rubin and T.L. Rauls, *J. Med. Chem.,* **19** (1976) 887.

13. P.G. Gassman, B.W. Cue Jr. and T.Y. Luh, *J. Org. Chem.,* **42** (1977) 1344.

14. G.R. Bedford and M.W. Partridge, *J. Chem. Soc.,* (1959) 1633.

15. L. Capuano and W. Ebner, *Chem. Ber.,* **104** (1971) 2221.

16. L. Heinisch and K. Kramarczyk, *J. Prakt. Chem.,* **314** (1972) 682.

17. F.J. Dicarlo and H.G. Lindwell, *J. Amer. Chem. Soc.,* **67** (1945) 199.

18. R.C. Elderfield and J.R. Wood, *J. Org. Ch.,* **27** (1962) 2463.

19. S. Somasekhara, V.S .Dighe, G.K. Suthar, S.L. Mukherjee, *current science.,* (1965)508.

20. W.C. Sumpter and W.F. Jones, *J. Amer. Chem. Soc.,* **65** (1943) 1802.

21. M.J. Taglianetti, *An. Fac. Farm. Odontol., Univ. Sao Paulo,* **7** (1949) 57; *Chem. Abstr.,* **45** (1951) 1997.

22. H. Cassebaum, *J. Prakt. Chem.*, **23** (1964) 301.

23. B.R. Baker, R.E. Schaub, J.p. Joseph, F.J. McEroy and J.H. Williams, *J. Org. Chem.*, **17**(1952) 164.

24. R.N. Castle, K. Adachi and W.D. Guither, *J. Heterocycl. Chem.*, **2** (1965) 459.

25. S.N. Pandeya, Siva Kumar Smith, Sridhar, *Acta Pharm.*, 55 (2005) 27.

26. R.K. Goyal, A.A .Mehta, R.Balaram, Elements of Pharmacology, Seventh Edition (2005)242.

27. www.epilepsy.com

28. Foye and O. William, Principles of Medicinal Chemistry, Fourth Edition (1995).

29. S. Louis Goodman and Alfred Gilman, The Pharmacological basis of Therapeutics, Vol. I, Eight Edition (1990) 302.

30. A. Remees William, In Wilson and Gisvold's , Text Book of Organic Medicinal and Pharmaceutical Chemistry, Ninth Edition 1991) 506

31. B.S. Chang, D.H.N. Lowenstein, , *Eng. J. Med.*, 349(13) (2003) 1257.

32. S. Louis Goodman and Alfred Gilman, The Pharmacological basis of Therapeutics, Vol. I, Eight Edition (1990) 522.

33. B.P Mallikarjuna, G.V.Suresh Kumar, B.S. Sastry, Nagaraj, Manohara K,P., *J. Zhejiang Univ Sci B.*, 8(7) (2007) 526.

34. www.medicinenet.com/epilepsysymptoms

35. G. Bertram, Katzung, Basic & Clinical Pharmacology, Eighth Edition (2001) 395

36. J.S. Duncan, *J. Clin. Pharmacol.*, 53(2) (2002) 123.

37. Derasari & Gandhi, Elements of Pharmacology, B.S. Shah Prakasan Publications, 14th edition (2005) 281.

38. H.P. Rang, Dell, M.N. and Ritter, J.M., Pharmacology Churchill Livingstone, New York, 4th Edn (2001) 566.

39. K.D. Tripathi, Elements of Medical Pharmacology, Jaypee Publications, 4th edition (2000) 384.

40. Shin-Hun Juang, Li-Rung Chen, Yu-Chin Wang, *Biorg & Med Chem Lett.,* 15 (2005) 3058.

41. Luhuia Lai, Lu Zhou, *J. Med. Chem.,* 49(2006) 3440.

42. A.K. Padhy, S.K. Sahu, P.K. Panda, D.M. Kar And P.K. Misro, *Indian J. Chem.,* 43B (2004) 971.

43. A. Raviraj, Kusanur, Manjunath Ghate And Manohar V. Kulkarni, *J. Chem. Sci.,* 116(5) (2004) 265.

44. S. Gupta, Raman, S.N.Vikas, Srivastava, *Asian J. Chem.,* 16(2) (2004) 779-783.

45. M. Ajitha, K. Rajnarayana And M. Sarangapani, *Pharmazie.,* 57(12) (2002) 796.

46. Kara L. Vine. *Biorg & Med Chem.,* 15(2007) 931.

47. Sriram, Tanushree, Balasubramani, *Biorg & Med Chem Lett.,* 15 (2005) 4452.

48. Klaus Kopka, Andreas Faust, Petra Keul, Stefan Wagner, Otmar Schober, *J. Med Chem.,* 49(2006) 6704.

49. V.Glover, J.M. Halket, P.J. Watkins, A. Clow, B.L. Goddwin and M.J. Sandler, *Neurochemistry.,* 51(1988)656.

50. J. Seidel and J. Wenzel, *Pol. Jr. Phamacol.,* 35(1979)407.

51. I.M .Mc Intyre and T.R Norman, *J. Neural Transm.,* 79 (1990)35.

52. S.K. Bhattarcharya and A.Chakraborti, *Indian. J. Exp. Biol.,* 36(1998)118.

53. S.K. Bhattarcharya , *Biog. Amines.,* 14 (1988) 131.

54. H.P Olesen and A. Kanstrup, Den. Pat. 97, 05,137, 13 feb 1997; ref .*Chem. Abstr.,* 126 (1997) 212050m.

55. Y. Evanno, M. Sevrin, C. Maloizel, O.Lgalbudec and P.George , *Chem. Abstr.,* 129 (1998) 282832h.

56. Y. Evanno, L.Dubois, M. Sevrin, F. Marguet and C.Gille, Pat. 9,906,406, 11 Feb 1999 ; ref . *Chem . Abstr.,* 130(1999) 168385f.

57. C. David T. Marie and G.Roussel, U.S.Pat.5,498,716, 12 Mar 1996 ; ref. *Chem. Abstr.,* 124 (1996) 34327r.

58. S. K. Srivastava, S. Srivastava and S. D. Srivastava, *Indian J. Chem.,* 38B (1999) 183-187.

59. G. S. Singh, T. Singh and R. Lakhan, , *Indian J. Chem.,* 36B (1997) 951-954.

60. S. N. Pandeya, I. Ponnilavarasan, A. Pandey, R. Lakhan and J. P. Stables, , *Pharmazie* 54 (1999) 12-16.

61. S. N. Pandeya, S. Smitha and J. P. Stables, , *Arch. Pharm. Pharm. Med. Chem.,* 4 (2002) 129-134.

62. S. N. Pandeya, A. Senthil Raja and J. P. Stables, *J. Pharm. Pharm. Sci.,* 5 (2002) 266-271.

63. M. Verma, S. N. Pandeya K. Singh and J. P. Stables, *Acta Pharm.* 54 (2004) 49-56.

64. G. Palit, R. Kumar, G. K. Patnaik and S. K. Bhattacharya, *Biogenic Amines* 13 (1997) 131-142.

65. A. E. Medvedev, A. Goodwin, A. Clow, J. Halket, V. Glover and M. Sandler, *Biochem & Pharmacol.,* 44 (1992) 5290-592.

66. L. Kennis, M. J. Edmund and C. Josephus, , Bel. Pat. 9,744,040, 27 Nov 1997; ref. *Chem. Abstr.,* 28 (1998) 34772e.

67. M. Sarangapani and G.Sammaaiah *Indian Drugs.,* 44(3) (2007) 200.

68. Maria Gline Mathews, *European J. Pharmacology.,* 556 (2007) 200.

69. Giscle Zapata, *Pharma Biochem & Beha.,* 86 (2007) 678.

70. D. Sasmal, D. M. Kar, L.Maharana, *Indian Drugs.,* 43 (9) (2006) 733.

71. R. S. Varma, R.K. Pandey and Piyush Kumar, *Indian J. Chem.,* 21(B) (1982) 775.

72. M. Sarangapani, Jessy Jacob, B. Srinivas and N. Raghunandan, *Indian Drugs.,* 38 (5) (2001) 264.

73. P.K Mukharjee, K. Saha, R.Balasubramanum, M.Pal, & B.P. Saha, *J. Ethnopharmacol.,* **54** (1996) 63.

74. R.A.Turner, Screening methods in Pharmacology, Academic Press, New York (1965) 72.

75. R.L. Krall, J.K. Penry, B.G. White, H.J. Kupferberg & E.A. Swinyard, *Epilepsia.,* 19 (1978) 409.